# TURKEY MOUNTAIN

## GUIDE TO THE GEOLOGY, FOSSILS, AND FEATURES OF UPPER TURKEY MOUNTAIN

STEVE KING

Photos and graphics by the author unless otherwise credited.

Graphics were constructed using data from the Oklahoma Geological Survey.

The walking route map was created from the trail map at River Park Authority.

Photo on Front Cover: Top of The Flight (Vcumketv)

ISBN: 978-1-962771-32-0
*Second Edition*

Library of Congress Control Number: 2026907994



Cover & Layout designed by Destini Richlin
*This book is set with typefaces Blue and Charter.*

Wise Life Books
Masterful Person Company Publishing
Stafford, VA
mpcpublishing.com

## WHAT TOUR PARTICIPANTS HAVE SAID...

"Wonderful information. I never knew Turkey Mountain had such a fascinating geology."

"Steve is an encyclopedia of knowledge about rocks, and it is interesting to hear and learn about a place I've visited hundreds of times and now know more about."

"It was incredibly informative and interesting!"

"Someone needed to do this."

# TABLE OF CONTENTS

# DISCLAIMER

This information is intended for education and informational purposes only. While every effort has been made to ensure the accuracy and reliability of the content, it is not a complete or thorough reference.

Use the guide at your own risk while hiking or biking. Be aware of potential hazards, including uneven terrain, wildlife and weather conditions. Exercise caution; stay alert, and take appropriate safety precautions.

Consult with qualified experts before making interpretations or decisions based on the information presented in this guide. The author disclaims any liability for errors, omissions or damages from using this guide.

Stay on the established trails. Everything described in this guide is on or along the trails in many places; therefore, no digging, moving or breaking of rocks is required. Most importantly, it is against park and city regulations.

## TULSA CODE OF ORDINANCES
## TITLE 26, SECTION 103

B. No person shall damage, cut, carve, mark, transplant or remove any plant or injure the bark; pick flowers or seed of any tree or plant; dig in or otherwise disturb grass areas; or in any other way injure the natural beauty or usefulness of any area.

C. No person shall cut any tree, dead or alive, whether erect or felled, or collect leaves, bark, wood chips, mulch or other organic material, stone, or soil for any purpose without the written permission of the appropriate Board or Director.

**$500 fine and up to 90 days in jail**

# PREFACE

I began leading impromptu geology tours in 2017 when my wife and I started mountain biking. She is also a geologist, so looking at rocks and interpreting their story is a typical outing.

Fast forward to 2019, during the final design phase of the Turkey Mountain Master Plan. The landscape architects sought a geologist who used Turkey Mountain, which coincided with the impromptu tours and mountain biking. We spent several hours walking the area north of the upper parking lot, discussing the geology, terrain and related topics.

Trail building under the master plan began in 2021, exposing more rocks, fossils and features. The tour evolved to incorporate new exposures and the stories they told. In 2022, the River Parks Authority added a small group geology tour to its offerings for a nominal fee. I began leading tours as a volunteer, with all proceeds benefiting River Parks. Interest exceeded expectations and positive feedback prompted the creation of the guide, providing information to more people at their convenience.

This guide presents a brief geological history to frame what we see today. It includes photos of common fossils and features on the trails, with concise descriptions avoiding lengthy technical discussions found in textbooks. Some features, like ripple marks, are unmistakable once their story is understood. They are seen in many locations on Turkey Mountain and throughout town on flagstone sidewalks and walls.

Rocks tell stories. Engage your imagination and visualize the environments described. A science background isn't necessary to enjoy your time on Turkey Mountain. Curiosity, a keen eye and this guide will enrich your experience.

This is a guide, not a textbook.

# ACKNOWLEDGMENTS

Thanks to my wife, Jeanette King, for the many hiking and biking outings to look at the rocks and interpret their stories. The discussions, debates and interpretations formed the baseline for the tour, and her attendance on many of the tours was always welcome. I'm looking forward to many more outings and discoveries.

I am grateful for River Parks Authority's providing the opportunity to lead small group geology tours and share the story of Turkey Mountain. I want to give special thanks to Ryan Howell at River Parks Authority. Ryan saw the tour's vision and suggested offering it to the public. His expertise and experience in the park, in scheduling, and participant registration were invaluable.

Many thanks to the individuals, science classes, Eco Club, geology club and hiking club who participated in the guided tours! Your interest, enthusiasm, insightful questions, and comments helped shape the tour.

# INTRODUCTION

A hidden gem of nature lies a short distance from downtown Tulsa, OK, amidst the urban sprawl. It's over 750 acres of urban wilderness within a series of parks and venues along the scenic Arkansas River managed by River Parks Authority. The appeal is the diverse terrain boasting over 45 miles of hiking and biking trails. The city's comforts and amenities are close, but it's large enough to escape the fast pace and noise to enjoy the serenity of nature.

Beneath the tranquil canopy is a rich collection of geologic events and wonders. Even a short hike on Turkey Mountain raises intriguing questions about its origin and history. How old is it? Why do we see so many rock layers? Do the rocks contain fossils? Was oil produced on Turkey Mountain? This guide answers these questions and many more.

We look to the rocks for answers. They are silent witnesses to local and global events, and they evoke the stories. Think of the rock formations as chapters in the story of Turkey Mountain, with the fossils and features found in the rocks as pages and illustrations.

The geologic story of Turkey Mountain was millions of years in the making. It's an interesting story involving tumultuous plate tectonic events, periods of mountain building, and a global ice age. These events created environments forming rock sequences in swamps, streams, rivers, flood plains, warm shallow seas, and deep seas. The rocks, fossils, and features in the rocks preserved the environment and narrative we interpret today.

The age of the rocks from the Arkansas River to the top of Turkey Mountain is about 305 million years ago (Mya) – from the late Carboniferous Period. This guide focuses on the types of fossils, and features in the rocks and sequence of processes that created them instead of extrapolating exact dates and timelines.

The story begins with what they are. The physical and chemical properties tell us the formations are sedimentary rocks. Micro to medium-grain shale, silt, and sand sediments were deposited in water environments, then lithified into rock layers. Plant fossils common to sandy, wet areas add to the story. Warm shallow seas created layers of fossiliferous limestone, while a deep-sea anoxic environment created black shale. Thin discontinuous layers of coal tell of wet, swampy environments. All are found from the river to the top of Turkey Mountain.

The rock layers changed many times in the +/- 300 vertical feet from the river to the top of Turkey Mountain. Each has an interesting narrative of creation and transition to the next layer. They're easy to identify with a keen eye and this guide. Use the guide to explore the trails of Turkey Mountain, identify the rocks and interpret the stories etched in the landscape.

Stay on the trails. No moving, digging or breaking of rocks and vegetation is necessary to see the rocks, fossils, and features described in this guide. Plus, and most importantly, it's prohibited by city and park regulations.

Be a good citizen and preserve the park for future enjoyment.

# HOW TO USE THIS GUIDE

Three options are presented for individual interests, abilities and time. Review the maps at riverparks.org. The QR codes on the trail markers link to an interactive map providing navigation and rescue points.

A. *Random hike*: Review the photos and information. Hike the trails, looking for the fossils and features described.

B. *Visit the natural stone staircase, The Flight (Vcumketv)*: The flight connects the lower and upper areas and contains most, but not all, fossils and features.

C. *Follow defined route:* The route with trail marker identification and GPS coordinates is for the more prominent fossils, features, and points of interest. The route begins at the upper parking area and concludes at Moonscape. Trail markers or GPS locations are provided with a note describing the discoveries at the stop, with photos and descriptions on the following pages. Seasonal vegetation, erosion and trail maintenance may obscure some fossils and features.

## MAP 1

1. Begin at the SE trailhead in the upper parking lot. (36.0703, -95.9936)
2. Proceed to the top of The Flight (Vcumketv) to see Calamites, Ripple Marks, and MISS. (36.0695, -95.9924)
3. Proceed east on Stay Gold (Hiking Only) a short distance, then north on the unimproved trail to an abandoned oil well and wellhead. (36.0691, -95.9903)
4. Return toward the parking lot to marker B41. Proceed north on West Gold Loop, (Nene Lane Herv).
5. Trace Fossils and Ripple Marks. (36.0735, -95.9930).
6. Proceed north on West Stay Gold Loop. (36.0755, -95.9934)
   Iron Banding and Liesegang Rings on the trail's east (right) side.
7. Continue to see MISS on the trail. (36.0753, -95.9920)

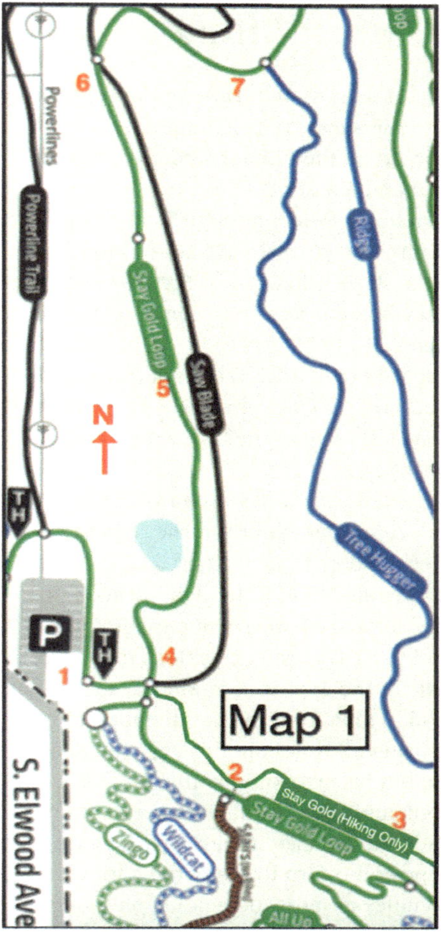
Powerlines
Powerline Trail
Stay Gold Loop
Saw Blade
Ridge
Tree Hugger
N
6
7
5
4
1
2
3
Map 1
Stay Gold (Hiking Only)
Stay Gold Loop
Zingo
Wildcat
All Up
S. Elwood Ave

## MAP 2

8. Stop at marker B44. Differential weathering is on the connecting trail to Ridge Trail.
9. Proceed up the unimproved connector trail to Ridge Trail and marker B62.
10. Turn north (left) on Ridge Trail and proceed to the feature known as Rock Garden, stopping at the top of the slope. (36.0786, -95.9919) See the photos of the erosion section to compare erosion contrast from 2017 to 2024.
11. Proceed down Rock Garden a few yards with caution. (36.0787, -95.9919) Large flat rock on the west (left) side of the trail with remnants of iron concretions.
12. Proceed down Rock Garden with caution to marker B63. A few yards north is the Coffeyville Shale and Checkerboard Lime contact.
13. Proceed north on Ridge Trail, walking on the Checkerboard Lime to Moonscap. (36.0817, -95.9922) The trail transitions to a black fissile shale at the approach to Moonscape. Moonscape is about a mile from the parking lot and is the end of the route. Several options are available for returning, but backtracking on this route is the most direct and least strenuous for an out-and-back hike. The total distance from the parking lot to Moonscape and return to the parking lot will be about 2.6 miles or more, depending on the return route.

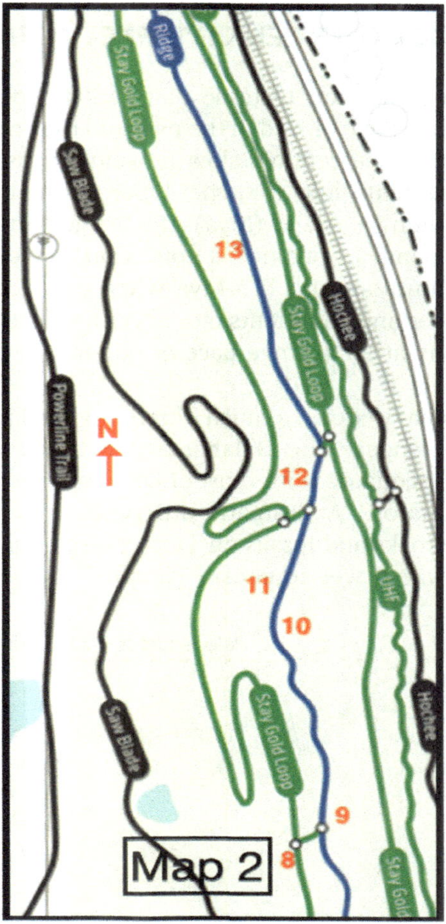
Ridge
Stay Gold Loop
Saw Blade
13
Hochee
Stay Gold Loop
Powerline Trail
N
12
11
UHF
10
Stay Gold Loop
Hochee
Saw Blade
9
8
Map 2
Stay Gold

## GEOLOGY OVERVIEW OF TURKEY MOUNTAIN

Continents are on tectonic plates that move over time. They have moved in the past and they continue to move. Imperceptible slow movement assembled the continents into a single supercontinent about 335 million years ago (Mya), the Pangea Supercontinent. Pangea remained a collection of close continents until around 175 Mya when it broke apart. The plates and continents moved to the locations we see today at an average pace of one inch per year.

Oklahoma occupied a much different global position during Pangea. The Oklahoma panhandle was on the paleoequator, with most of Oklahoma below the paleoequator.[1] A dramatic difference from today, but the rocks and fossils on Turkey Mountain represent a warm, wet equatorial climate.

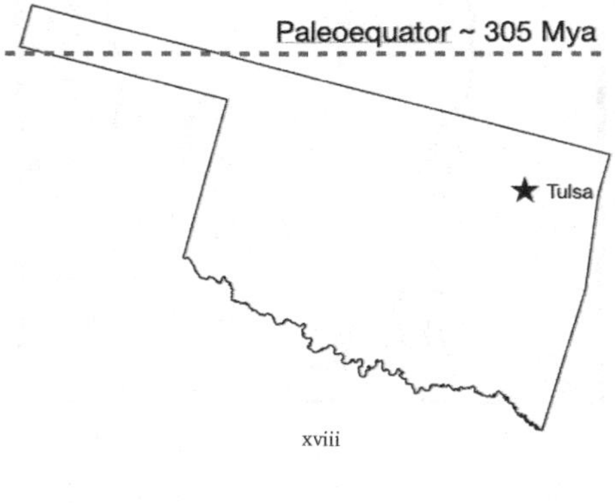

Sedimentary rocks compose the visible +/- 300 vertical feet of Turkey Mountain from the Arkansas River to the upper parking lot. The age is the late Carboniferous Period at about 305 Mya. Five named formations comprise the +/- 300 ft of rock. The Memorial (~50 ft) is near the river, progressing up through the Lost Branch (~40 - 65 ft), Seminole (~85 ft) Checkerboard Lime (~3 ft), and Coffeyville (~95 - 120 ft) as the uppermost. An unconformity or erosional contact between the Lost Branch and Seminole accounts for the Lost Branch range of thickness. Erosion of the Coffeyville accounts for the range of thickness.[2]

This guide focuses on Coffeyville and Checkerboard's most common fossils and features. Coffeyville is the most visible and accessible from the upper parking lot, with the older thin Checkerboard Lime below it.

**Turkey Mountain Formations and Thickness**

| | |
|---|---|
| Pcv | Coffeyville, ~ 95 - 120 ft |
| Pcb | Checkerboard Lime, ~ 3 ft, |
| Pam | Seminole, ~ 85 ft |
| Unconformity | |
| Plo | Lost Branch, 40 - 65 ft |
| Mmm | Memorial, ~ 50 ft |

# NOTES

# THE CARBONIFEROUS PERIOD (359 - 299 MYA)

## An overview

Carboniferous translates to coal-forming or carbon-bearing. Plants absorb carbon dioxide, expel oxygen, and use the carbon to create and store energy. Swamps covered broad areas, sequestering vegetation and stored carbon in a low oxygen environment, forming peat. Continued burial produced pressure and heat. The pressure and heat converted the peat to extensive coal deposits, thus the name Carboniferous. Major deposits of Carboniferous coal stretch from the eastern US, including Oklahoma, through Europe and China.

Discussion of Oklahoma geology often includes a reference to being underwater. It's frequently described as a single event, but multiple cycles of below and above sea level occurred with fluctuations between the extremes. Some events below and above sea level spanned millions of years, while others were shorter. The main forces of sea level change are plate tectonics, ice ages and mountain-building events.[3]

The Karoo Ice Age[4] was the fourth of five major ice ages in the past 2.4 billion years. It began at about 360 Mya and ended at around 255 Mya. An ice age is the period from the first glaciation to the final warming. Between the first and final events, glaciation and interglacial warming cycles occurred while the equatorial climate remained stable, much like today. About 305 Mya, the visible Turkey Mountain formed about the midpoint of the Karoo timeline.

Glaciation lowered the sea level as snow and ice accumulated as polar ice sheets, and glaciers at high elevations. The opposite occurred during interglacial warming periods. Sea level rose as glaciers and polar ice cover melted. Sea level fluctuations of hundreds of feet were common between the extremes.

Each sea level change produced different conditions for creating or removing rock layers. Higher sea levels produced conditions for limestones such as the Checkerboard Lime during warm periods when the sea level was higher. Streams, rivers and flood plains created siltstones and sandstones during periods of glaciation when sea levels were lower. Offshore marine shales often formed during the transition between lower and higher sea levels. The most visible shale, siltstone and sandstone sequences are the

Coffeyville Formation at the upper parking lot. Sea level changes also formed the older formations of Seminole, Lost Branch, and Memorial, below the Checkerboard but are less visible and accessible.

A series of mountain-building events also created and shaped the Oklahoma landscape. Global plate tectonics pushed the South American plate into the North American plate. The collision pushed sedimentary layers north into thrust faults. The uplift and folding created the Ouachita mountains in SE Oklahoma and SW Arkansas. Pressure from the closing plates triggered additional uplifts and mountain building. The Nemaha ridge in central OK uplifted. The Arbuckle Mountains formed and would have towered higher than the Rocky Mountains had they not eroded. The final Oklahoma mountain-building event created the Wichita Mountains.[4]

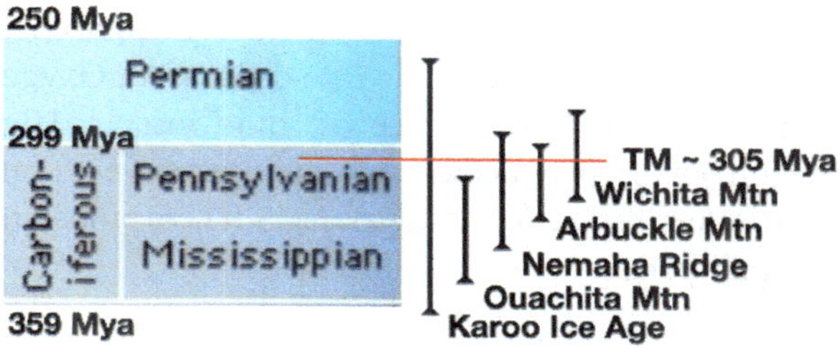

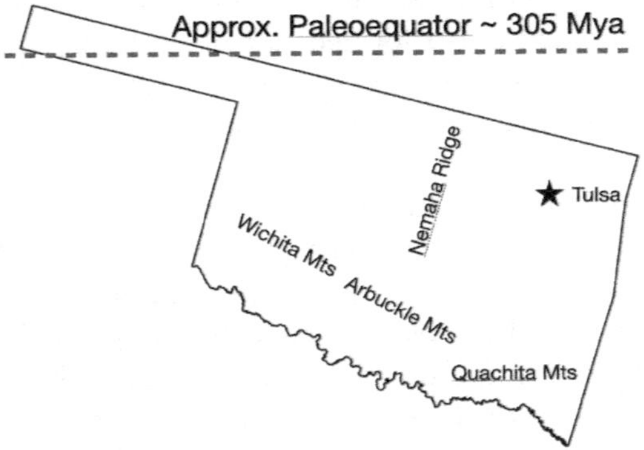


Vegetation was prolific in swampy areas with a warm, wet, perpetual growing season.[5] Vegetation consumed carbon dioxide and expelled oxygen to the atmosphere. Sequestered carbon in the swamps removed it from the carbon cycle; thus, carbon dioxide declined from around 1,000 ppm to less than 200 ppm during the Carboniferous Period.[6] Oxygen in the atmosphere increased to 35% compared to 21% today, while carbon dioxide decreased. Oxygen availability governs the size of some insects. High oxygen levels produced colossal insects, including dragonflies with 2.5 ft wingspans, 3 ft scorpions and 8 ft millipedes.

In short, the Carboniferous Period was unique in Earth's history. Plate tectonics positioned Oklahoma on the equator and produced mountains and extensive swamps. An ice age raised and lowered sea levels. The fluctuations created a range of environments and sedimentary formations. The earth recovered from an 80% mass extinction 70 million years earlier with an explosion of vegetation. The extensive swamps sequestered prolific vegetation, producing a massive amount of coal. It's also the only period when temperatures and carbon dioxide levels were similar to today in the past 600 million years.

## IRON

A short walk on the trails reveals a range of earth-tone rocks from almost white to brown and yellow to dark red. Some are uniform in color, while others vary to extremes over a short distance.

Iron is the fourth most common element in the crust; less than 6% behind oxygen, silicon, and aluminum, yet it influences the color of sedimentary rocks. Dissolved iron at 0.3 to 2.0 ppm is typical in untreated fresh water. Water remains colorless, while iron remains in solution as ferrous (II) iron. Changes in saturation, pH, and oxygen availability oxidize and precipitate ferric (III) iron. The most common iron-rich minerals are hematite, limonite, and goethite. They stain the individual grains and form iron minerals within the rock matrix.

Hematite ($Fe_2O_3$) is red and indicative of high-oxygen environments such as rivers and flood plains. Limonite ($FeO(OH) \cdot nH_2O$) is yellow, and Goethite ($FeO(OH)$) is brown. Each represents decreasing amounts of available oxygen, but usually terrestrial. Ferrous iron (II) is less common, producing a yellow-green color. It forms in shallow marine environments with less oxygen than in rivers and floodplains.[7]

Iron oxides are common, non-reactive and non-toxic in small quantities. They began being used in ancient times as pigments in paint. Hematite was the most common pigment used in the earliest cave paintings. Common pigment applications also include cosmetics, textiles and glass.

## CALAMITE

Calamites spanned the Carboniferous and Permian periods. They were an ancestor to modern-day Horsetails but resembled trees reaching heights up to 60 ft without having a woody composition. The interior was pithy, resembling tall grasses. The exterior had vertical ribbing with regular spaced joints or nodes. The trunk and limbs had a small hollow center, like tall grasses.[8]

A continuous equatorial growing season in sandy and swampy environments spurred prolific growth. The result was calamites became the underpinning of many carboniferous coal beds. The beds stretch from the eastern US, including Oklahoma, to Europe and Asia.[9]

Calamite fossils are common in the upper Coffeyville. The natural rock staircase between the lower and upper parking lots has distinct Calamites. Other specimens are throughout the park.

Calamites

Calamites

# LYCOPHYTE

The most common arborescent (tree-like) Lycophytes were Sigillaria and Lepidodendron. Both are related to modern club mosses, with Sigillaria maturing to about 90 ft. and Lepidodendron maturing to about 100 ft. The trunk exterior resembled scales from leaf scars instead of typical bark.

Both are common Pennsylvanian-age fossils found in Oklahoma.[8] They grew in wet, swampy areas, with Sigillaria growing in drier areas of the swamp. Lepidodendrons grew in wetter but not the wettest areas. These lycophytes became extinct when the swamps disappeared during the Permian period.

Stigmaria is the root system of arborescent lycophytes often found as fossilized cylinders. The outside has small round dimples spiraling around the outside. The dimples have a small bump in the center of better-preserved specimens. Extraordinary specimens have a small fossilized root present in place of the bump.

A fossilized Lepidodendron stump and a Stigmaria root system have been identified in separate locations.

Stigmaria (Lycophyte Root System)

Lepidodendron Stump

# NOTES

## RIPPLE MARKS

Ripple marks are sedimentary structures formed in loose mud, silt, or sand by moving water or wind. They're common in streams, rivers, beaches and arid settings and are identified as a series of elongated parallel crests and troughs. The ripple mark categories are symmetric and asymmetric.

The back-and-forth (bidirectional) motion of waves on a beach creates symmetric ripple marks. They form parallel rows of crests and troughs perpendicular to the wave direction. Each side of the ripple has an equal length and slope between each trough and crest.

Water flowing in one direction, like streams and rivers, creates ripple marks. A disturbance in water flow over or around an irregularity, such as a rock or branch, induces ripples or waves. Narrowing of a flow channel forces more water into less space, also inducing ripples or waves. Surface ripples produce sediment ripples, which tend to cause more ripples downstream. These are asymmetric, having sides of unequal length and slope perpendicular to the flow direction.

The long side of each asymmetric ripple mark is the stoss side, and the short steep side is the lee side. Water flow moves sediments up the long side, and they fall to the shorter lee side. Thus, the upward direction of the stoss (long) side is the direction of water flow. The steep side is always downstream.

Ripple marks are common on the siltstone and sandstone layers. They vary from a few millimeters to several centimeters in height from the base of the trough to the peak of the crest. They are also seen in other locations on thin flagstone pavers on sidewalks and walls.

# TRACE FOSSILS

Trace fossils, or Ichnofossils, are traces of ancient animal activity. They include tracks, burrows and tunnels, but not the animal. They identify the re-working of sediments before becoming part of the rock record. Scavengers and decay often consume the animals before fossilization.

The most common trace fossils on Turkey Mountain are burrow entrances on the surface of flat rocks. They are often only a few millimeters in diameter, with larger diameters being less common. They appear as groups of small dimples or indentions on flat rocks. Trails between burrows are sometimes present with irregular patterns. Some trails could have originated as tunnels that collapsed before the rock lithified. Other trails may have been tunnels exposed as the top layer of rock eroded, exposing the tunnels. (36.0735, -95.9930)

Trace Fossils
Tunnels in Broken Section
Burrows
GSA
IN
CM
mm
Grain Size Scale

## LIESEGANG BANDS AND RINGS

Liesegang bands are yellow to red bands that form in siltstone and sandstone. The bands are perpendicular to the bedding plane, while the rings are circular or irregular. Space between the grains is porosity, while permeability is the ability to move fluid through the porosity. Both are necessary properties to form Liesegang bands and rings.

Dissolved iron at 0.3 to 2.0 ppm is typical in untreated water. Water with dissolved iron moves through the rock over time. Water remains colorless, while iron remains in solution. Changes in saturation, pH and oxygen availability oxidize and precipitate the iron. Precipitated iron stains the grains and occupies the space between grains, creating the feature.

Liesegang bands/rings are created by horizontal flow parallel to the bedding plane. Uniform rock properties create uniform repeating Liesegang bands or rings. The converse is also true; inconsistent properties create irregular Liesegang bands or rings. Vertical flow through the rock creates banding parallel to the bedding plane. These are iron band-

erva
es be
ognitio
om simila
s. MISS fea
as algae, bac
nities producin
on or within sed
nderstanding the en
erved.[11]

uce Extracellular Polyme
hich create biofilms.[12] Bi
e agent for microbes to for
act as adhesives, anchoring an
s to surfaces, plus providing a pr
against toxins and minor damag
sediments on bedding planes, the mo
MISS feature on Turkey Mountain. Th
n or build up over time, creating thick la
mounds. Their appearance is best describ
nkles on the rock surface.



ng, not Liesegang bands. Some rocks combine Liesegang bands and iron banding, creating intricate patterns.[10] (36.0755, -95.9934)

Beverage coasters crafted from porous sandstone with interesting Liesegang patterns and banding are common. The decorative, porous natural stone coasters absorb moisture and provide a conversation piece.

Iron Banding & Liesegang Bands/Rings

## MICROBIAL INDUCED SEDIMENTARY STRUCTURES (MISS)

MISS became a new classification of sediment
rock features a few decades ago, although obs
tion, reporting, and speculation began deca
fore the classification. New studies led to rec
as a distinct category to distinguish it fr
structures created by geologic processe
tures result from microorganisms such
teria and other microbial commu
biofilms. The structures appear
ments, which are essential to u
vironment of deposition pre

Colonizing microbes prod
ic Substances (EPS),
films act as a cohesi
colonies. They also
stabilizing coloni
tective coating
Biomats bind
recognized
can be thi
ers and
as w

ing, not Liesegang bands. Some rocks combine Liesegang bands and iron banding, creating intricate patterns.[10] (36.0755, -95.9934)

Beverage coasters crafted from porous sandstone with interesting Liesegang patterns and banding are common. The decorative, porous natural stone coasters absorb moisture and provide a conversation piece.

Iron Banding & Liesegang Bands/Rings

Liesegang Bands
IN
CM

CM
IN
Iron Banding

Sandstone Beverage Coaster with Iron Banding
METRIC 1 2 3 4 5 6 7 8 9

## MICROBIAL INDUCED SEDIMENTARY STRUCTURES (MISS)

MISS became a new classification of sedimentary rock features a few decades ago, although observation, reporting, and speculation began decades before the classification. New studies led to recognition as a distinct category to distinguish it from similar structures created by geologic processes. MISS features result from microorganisms such as algae, bacteria and other microbial communities producing biofilms. The structures appear on or within sediments, which are essential to understanding the environment of deposition preserved.[11]

Colonizing microbes produce Extracellular Polymeric Substances (EPS), which create biofilms.[12] Biofilms act as a cohesive agent for microbes to form colonies. They also act as adhesives, anchoring and stabilizing colonies to surfaces, plus providing a protective coating against toxins and minor damage. Biomats bind sediments on bedding planes, the most recognized MISS feature on Turkey Mountain. They can be thin or build up over time, creating thick layers and mounds. Their appearance is best described as wrinkles on the rock surface.

Trace fossils are linked to MISS. The animals secrete a biofilm as they create burrows, tracks and tunnels.

## DIFFERENTIAL WEATHERING

A term describing irregular deterioration of exposed surfaces. Water, wind, sunlight, and plants are the most common weathering forces. Weathering breaks down or dissolves the rock, while erosion transports the material away. Variations in rock mineralogy and cementing material determine the resistance to weathering. Variations create irregular surfaces, while uniform properties create smooth, uniform, weathered surfaces. Hematite laminations and Liesegang bands are more resistant to weathering, creating differential weathering. (B44, 36.0766, -95.9917)

Differential Weathering

# EROSION

Erosion is an ongoing natural process on exposed outdoor natural surfaces. The original trails are best described as ad hoc trails created to get from point A to Point B. They were made without regard to erosion, sustainability, or future use. Some were roads or paths to early homesteads, while others were roads and paths from the early oilfield days a century earlier. Erosion on Ridge and Powerline Trails are examples of advanced erosion.

The Turkey Mountain Master Plan includes erosion management with sustainable trails. The new trails have slope and contour parameters that direct rain runoff away from the trail, reducing rain puddling and thus promoting faster drying. The sustainable trails reduce pedestrian and bicycle traffic on older ad hoc trails, which are more prone to erosion, reducing the need to create prohibited ad hoc trails.

The Rock Garden section of Ridge Trail is a popular hiking and biking area toward the north end of Ridge Trail above the junction of Ridge, Saw Blade and Stay Gold.

It's a collection of small to large rocks on a long moderate slope, creating a technical challenge for mountain bikers. It's an example of uncontrolled erosion on a bare trail with heavy traffic. This rocky slope is on the lower unit of the Coffeyville formation, which is shale. The shale is more susceptible to erosion. The harder siltstone and sandstone remnants of the upper layers settle lower as the shale erodes from around the rocks.

Minor to extensive erosion occurs on many slopes, trails and ravines. The photos show erosion on Rock Garden over seven years without mitigation or remediation. (36.0786, -95.9920)

2017- Rock Garden

2024 - Rock Garden

# SHALE

Shale is the most common sedimentary rock, with clays as its main constituent. A low percentage of very fine-grained quartz silt at 8 microns (0.008 mm) or less is typical. Minor amounts of other minerals may also be present. It's ordinary mud before compaction and lithification; therefore, it's sometimes called mudstone when it lacks laminations or layering.

Clay is a mineral classification. The most common are kaolinite, illite, chlorite, and smectite. The chemical composition is silica, alumina or magnesium oxide, and water, forming flat sheets or plate-like structures. Laminated, flaky or fissile shale reflects the sheet-like chemical structure of clays. Clay also defines grain size as 2 microns (0.002 mm) or less, about 1/25th the width of a human hair. At 2 microns, clay particles are not visible with a 10x field hand lens, so complex tests are necessary.[13]

Clear-running streams and rivers often turn red or brown with suspended clays and silt in fast-moving water. When the energy and turbulence of moving water decline, suspended particles settle to the bottom. The smallest particles are often transported farther from shore into deeper water.

Shale deposits in oxygen-rich environments derive color from a low percentage of iron oxide and iron hydroxides. Ferric iron compounds yield red, yellow, and brown shale, while ferrous iron yields greenish shale. A low percentage of unoxidized organic content yields black shale in an anoxic environment.

Clays can absorb water and exchange ions, making them useful for various applications. A catalyst in chemical reactions and removing contaminants in water are among the uses. Shale (clay) has a plastic consistency when mixed with water and becomes solid when dry. This property is ideal for making ceramics with the oldest known clay pots dating back twenty thousand years. Other common uses of clays and shale include producing cement, bricks, tile, and pots, plus it's often a hydrocarbon source rock.

The lower unit of the Coffeyville Formation is notable. Frankoma Pottery moved to Sapulpa, OK, in 1954 and began producing pottery from the Coffeyville Shale.[14] The type of clay, clay-to-silt ratio, color, and other subtle properties comprise a good base for pottery.

## CONCRETIONS

Concretions are spheres, ovoids and sometimes irregular shapes of cemented sediments within a body of sedimentary rock. They begin with the same silt/ sand mineralogy as the host rock. The typical size is often 0.5 inches up to a few inches, but larger ones of many yards are possible. They occur in siltstones and sandstones such as the ones on Turkey Mountain.

They begin with the quick burial of a plant or animal.Something as simple as a leaf or bone can be a nucleus for creating a concretion. Decay begins soon after burial before the rock lithifies, releasing minerals that combine with minerals in the water. They precipitate as a cementing agent, bonding the silt or sand grains. The original thinking was they formed over hundreds or thousands of years. New investigations show some, but not all, mineral combinations may precipitate in weeks, months or a few years.[18] This varies with the cementing agent under ideal conditions. The cementing material forming a concretion is more weather-resistant than the surrounding rock. Therefore, concretions are often seen as convex bumps on rocks or weathered free from the rock as smooth stones or nodules. Some contain fossils, but most organic nuclei decompose before fossilization occurs.

Iron concretions begin as simple concretions. They become rust-colored iron concretions when surrounded with an iron oxide (hematite, $Fe_2O_3$) rind. The inside remains the same silt/sand mineralogy as the surrounding host rock outside the rind.

The more weather-resistant hematite rind creates differential weathering. The softer rock matrix weathers, exposing the dark red convex surface of the rind. Further weathering will free the iron concretion from the rock or leave a concave cup-shaped rind on the rock surface. (36.0787, -95.9918)

## CHECKERBOARD LIME

The Checkerboard Lime is below the Coffeyville and above the Seminole formation.[2] It's a medium gray limestone with a low percentage of silica and clay, sometimes called a carbonate mudstone. It's +/- 3 ft thick in this area, fracturing into large rectangular blocks, thus called Checkerboard. Fresh surfaces are the typical gray (calcium carbonate) limestone color but weather to yellow-orange to yellow-brown over time. Thin calcite veins are present in some exposures. Consistent thickness and a tendency to fracture in rectangular blocks make it well suited for traffic barriers and landscaping.

The limestone is a product of a warm shallow sea environment because of its proximity to the equator and being shallow; represented by the type of rock (calcium carbonate) and marine fossils. The thin layer is continuous but less visible because of soil and vegetation cover. However, it is well exposed at the bottom of Rock Garden on the northern end at the junction of Ridge, Saw Blade and Stay Gold trails. (approx. 36.0794, -95.9916) Small blocks of Checkerboard are seen along the lower Stay Gold Loop as displaced downslope remnants. Small blocks are also seen on the staircase as border columns about midway.

Whole marine fossils and fossil debris are major constituents of the Checkerboard. Bivalves (pelecypods), brachiopods, bryozoans, coral, and crinoids are the most common, with smaller fossils seen with a 10x hand lens.[15] It's rare to find individual whole fossils weathered from the rock, but partial exposers on the weathered rock surfaces, such as the boulders along Elwood Avenue at the upper parking lot, are easy to see.

IN
mm
Checkerboard Lime

## BLACK FISSILE SHALE

This shale is most visible in an area known as Moonscape (36.0817, -95.9922) toward the north end of Ridge Trail. The location is popular with hikers and bikers for its smooth surface and gentle slopes. The 5- to 10-foot-thick black shale is part of the lower unit of the Coffeyville, sitting on the Checkerboard Lime. Ridge Trail runs between large car-size sandstone boulders on the black shale.

Black shales form in low oxygen (anoxic) deep-water marine environments. The predominant shale mineral is clay, with a flat sheet-like structure reflected in parallel laminations. The dark gray to black shale derives color from a low percentage of carbonaceous material plus 1% to 5% phosphate.[16] Fissile describes breaking or splitting into thin parallel layers with flat surfaces. Phosphate nodules are sometimes present as spheres, ovoids, and irregular hard nodules. Most are an inch or less in diameter and contain up to 30% phosphate.

Phosphate in the shale originates from weathering and erosion of phosphate-bearing minerals such as apatite. Rivers carry it to the ocean, where it becomes a nutrient available for plants and animals. Algae and phytoplankton use phosphate to synthesize biological molecules and sometimes their shells. Phosphatic compounds are also a component of vertebrate bones.

Algae, phytoplankton, and other marine life settle on the seafloor when they die. Organic matter decomposition is slow in an anoxic environment, concentrating the carbon and phosphate. Additional sediments bury the carbonaceous material, preventing oxidation.

Some black shales are mined for phosphate used as fertilizer and industrial applications. Black shale may also contain commercial quantities of cobalt, copper, nickel, and other valuable metals used in the EV battery market. When buried and heated over enough time, organic-rich shales are source rocks for hydrocarbons.

Black Fissile Shale

Black Fissile Shale at Moonscape

## CHANNELS AND STACKED CHANNELS

The upper Coffeyville formation consists of stream, river, and floodplain deposits. Streams and rivers meander, reshaping their floodplains during fast moving floods, often creating new channels while filling old ones. Single and overlapping or stacked channels form over time, becoming preserved in the rock record when additional sediments fill and bury the channel(s).

Channels are a few feet to hundreds of feet wide. They are easy to identify on cliffs or bare rock faces. Upturned channel edges have a smiley face as their signature. Only one end may be visible due to soil, vegetation, and rock debris covering the other end.

The east and south sides of Turkey Mountain have the most vertical rock faces. Visible channel deposits (smiley faces) are also on other trails with the best viewing in winter after the leaves fall.

Stacked Channels

## CREEP

Creep is the imperceptible slow movement of soil and surface material downslope. Trees are the most common and interesting evidence of creep on Turkey Mountain. A curved trunk near ground level indicates the tree moved downslope during its youth. Over time, the growth direction corrected to vertical, thus creating the curve.

Frost and freeze heaving are common methods of creating creep. When wet soil freezes, it expands perpendicular to the slope. The expanded frozen material then drops straight down upon thawing. Freeze-thaw cycles on a slope create an imperceptible small zip-zag or sawtooth movement downslope over time. Creep is sometimes reflected in tilting telephone poles, fence posts, and structures.

Creep

## SPRINGS

A spring is a natural exit of groundwater seeping, dripping or flowing from underground. Shales are often involved in the process because they exhibit very low permeability, hindering water movement. Surface water filters down through soil, rock, and cracks until stopped by an impenetrable barrier, which is often shale. Gravity continues to pull the water downslope along the impermeable boundary until it reaches an exit point such as a hillside, ravine or road cut, creating a natural spring. It can be short-lived or continuous, dependent on the nature of rocks upslope and rainfall.

The rock formations of Turkey Mountain dip to the west at about 60 ft per mile. Water seeps from some slopes on Powerline Trail. Minor water seepage is also seen on a few slopes and ravines further west. Due to the slow gravity-fed water migration, ravines often stay damp to muddy long after rainfall stops.

## OIL PRODUCTION ON TURKEY MOUNTAIN

The Glenpool oil discovery south of Turkey Mountain was a worldwide headline in 1905. This pre-statehood discovery sparked an oil boom with hundreds of wells drilled in the Turkey Mountain district. Most of the oil was produced during a surge of drilling activity from the 1910s through the 1920s.[17]

Several oil-producing formations range from ~450 ft to about ~2,200 ft below the surface. Some produced hydrocarbons from a single formation, while others produced from multiple formations. Most had low daily production but were economical for decades. Wells were abandoned decades ago when production became uneconomical. A few abandoned wells are identified by the 4" to 6" vertical steel pipe/casing extending above ground, with a well retaining the wellhead equipment at 36.0691, -95.9903. A few concrete supports for pump jacks and misc. oilfield equipment are still in place at several locations along the trails. Discarded coils of steel cable are present with a few 2" pipes that carried production from the wellhead to storage tanks on the surface.

The Tulsa Geological Society gifted the people of Tulsa County a granite monument in 1985. The monument features a simple oil pool map of Tulsa County on the top with a chronology of significant oil and gas events on the sides. It is next to the paved pedestrian/bike trail at 68th and Riverside. A vertical pipe in the river SW of this location, about 50 ft from the east riverbank, is an abandoned well. Others dot the riverbed north and south of this point.

Abandoned oil well with wellhead equipment

Donated by Tulsa Geological Society

## SUMMARY

Turkey Mountain is a convenient and serene escape to the Carboniferous period when Oklahoma was part of the Pangea Supercontinent. It's a brief chapter on the earth's fascinating history, including the rise and fall of sea levels from plate tectonics, an ice age, and mountain building that created the rock formations from the river to the upper parking lot.

The fossil record shows that diverse environments were home to a variety of life. Calamites are a snapshot of swampy areas, while fossil debris of crinoid stems, brachiopods, coral, and others thrived in a warm shallow sea. In contrast, a black fissile shale created in a static, anoxic deep-sea environment is void of fossils.

The Coffeyville and Checkerboard Lime are the youngest and most accessible formations. They are featured in this guide as upper Turkey Mountain. A keen eye for detail, plus this guide provides information and photos to identify the most common fossils and features. A science background isn't necessary to enjoy the discovery of fossils and features created during this period.

## TULSA CODE OF ORDINANCES
## TITLE 26, SECTION 103

B. No person shall damage, cut, carve, mark, transplant or remove any plant or injure the bark; pick flowers or seed of any tree or plant; dig in or otherwise disturb grass areas; or in any other way injure the natural beauty or usefulness of any area.

C. No person shall cut any tree, dead or alive, whether erect or felled, or collect leaves, bark, wood chips, mulch or other organic material, stone, or soil for any purpose without the written permission of the appropriate Board or Director.

**$500 fine and up to 90 days in jail**

# REFERENCES


1. *Late Pennsylvanian Paleography of Oklahoma*
   Oklahoma Geological Survey, ogs.ou.edu/earthscience/intgeol/6-Paleo.Upper%20Penn.jpg

2. *Geologic Map of the Jenks 7.5' Quadrangle*
   T. M. Stanley & J. M. Chang
   Tulsa County, Oklahoma Geological Survey, ogs.ou.edu/docs/OGQ/OGQ-85-color.pdf

3. *Geologic Time Scale*
   Oklahoma Geological Survey, ou.edu/content/dam/ogs/images/OGS-OK_geologic_time_scale.png

4. *Glad You Asked: Ice Ages – What Are They And What Causes Them?* S. Eldredge & B. Biek, geology.utah.gov/map-pub/survey-notes/glad-you-asked/ice-ages-what-are-they-and-what-causes-them/

5. *The Carboniferous Period,* Laura Klappenbach, thoughtco.com/carboniferous-period-129666

6. *Better Graph of CO2,* G. Horner
   medium.com/@ghornerhb/heres-a-better-graph-of-co2-and-temperaturefor-the-last-600-million-years-f83169a68046

7. *Goethite,* Encyclopedia Britannica, britannica.com/science/goethite

8. *Common Fossils of Oklahoma*, Sam Noble Museum, samnoblemuseum.ou.edu/common-fossils-of-oklahoma/

9. *Fossil of the Month: Calamites*, Kentucky Geological Survey, uky.edu/KGS/fossils/fossil-month-12-2021-calamites.php

10. *Liesgang Bands & Rings*, Wikipedia en.wikipedia.org/wiki/Liesegang_rings_(geology)

11. *Microbially Induced Sedimentary Structures,* Nora Noffke, Encyclopedia of Astrobiology, doi.org/10.1007/978-3-642-27833-4_1004-4

12. *Microbial Extracellular Polymeric Substances (EPS) in Fresh Water Sediment,* Sabine Ulrike Gerbersdorf, Bernhard Westrich, & David M. Paterson, Microbiology of Aquatic Systems, doi.org/10.1007/s00248-009-9498-8

13. *Field Identification And In-Situ Testing,* Indiana Department of Transportation, in.gov/indot/engineering/files/Chapter-4-Field-Identification-and-In-situ-Testing.pdf

14. *Roadside Geology of Oklahoma*, Neil H. Suneson, Mountain Press Publishing Company, pg 118

15. *Pennsylvanian Atlas of Ancient Life Midcontinent United States*, pennsylvanianatlas.org/species/

16. *Petrology and Origins of Phosphate Nodules From the Midcontinent Pennsylvanian Epicontinental Sea*, David L. Kidder, Journal of Sedimentary Petrology, Vol 55, No.6, Nov, 1985, p. 809-816

17. *Geology and Mineral Resources of Tulsa County Oklahoma*, Malcolm C. Oakes, Oklahoma Geological Survey, Bulletin No 69, pg 123-124, ogs.ou.edu/docs/bulletins/B69.pdf

18. *Scientists Have Cracked the Mystery of How Fossil Concretions Form*, David Bressan, Forbes, May 3rd, 2018, forbes.com/sites/davidbressan/2018/05/03/scientists-crack-mystery-of-how-concretions-form/

# BIOGRAPHY

Steve King is a retired geologist in Tulsa, Oklahoma, with over 40 years in the oil and gas industry. A graduate of Oklahoma State University, he specialized in enhancing oil and gas production.

In retirement, he created and leads a geology hike for non-geologists as a volunteer. He has presented talks for the library's Summer Reading Program and for local clubs, translating complex geology into clear, everyday language for general audiences.

Steve enjoys bicycling, hiking, reading, and traveling. He chuckles when he says vacations sometimes resemble field trips.

Contact: steveking5139@gmail.com

Instagram: steveking5139

Podcast: rsuradio.com/podcast/10338/

Made in United States
Orlando, FL
15 July 2026

84286103R00046